Theodore Gill

Arrangement of the Families of Mollusks

Theodore Gill

Arrangement of the Families of Mollusks

ISBN/EAN: 9783337864958

Printed in Europe, USA, Canada, Australia, Japan

Cover: Foto ©berggeist007 / pixelio.de

More available books at **www.hansebooks.com**

SMITHSONIAN MISCELLANEOUS COLLECTIONS.

227

ARRANGEMENT

OF THE

FAMILIES OF MOLLUSKS.

PREPARED FOR THE SMITHSONIAN INSTITUTION

BY

THEODORE GILL, M. D., Ph. D.

WASHINGTON:
PUBLISHED BY THE SMITHSONIAN INSTITUTION,
FEBRUARY, 1871.

ADVERTISEMENT.

THE following list has been prepared by Dr. Theodore Gill, at the request of the Smithsonian Institution, for the purpose of facilitating the arrangement and classification of the Mollusks and Shells of the National Museum; and as frequent applications for such a list have been received by the Institution, it has been thought advisable to publish it for more extended use.

JOSEPH HENRY,
Secretary S. I.

SMITHSONIAN INSTITUTION,
WASHINGTON, January, 1871

ACCEPTED FOR PUBLICATION, FEBRUARY 28, 1870.

(iii)

CONTENTS.

(v)

INTRODUCTION.

OBJECTS.

THE want of a complete and consistent list of the principal
subdivisions of the mollusks having been experienced for some
time, and such a list being at length imperatively needed for the
arrangement of the collections of the Smithsonian Institution, the
present arrangement has been compiled for that purpose. It must
be considered simply as a provisional list, embracing the results
of the most recent and approved researches into the systematic
relations and anatomy of those animals, but from which innovations and peculiar views, affecting materially the classification,
have been excluded. The only merit which is claimed for it is
the embodiment and co-ordination, it is hoped in a tolerably consistent form, of the taxonomic results of the information scattered
through many volumes. There will doubtless be much diversity
of opinion respecting the relative value of certain groups, as well
as of the characters themselves whose modifications have been
used for the limitations of the groups, and the author will not
disguise that he himself entertains much doubt respecting certain
groups and relationships preserved in the arrangement. It has
seemed advisable, however, to provisionally adopt the opinions
of those who have most thoroughly investigated the different
groups rather than to introduce innovations based on hypothe-
tical considerations, and which would be perhaps found to be
liable to as many objections as those adopted

But although, from the very nature and extent of the subject,
the present arrangement is a compilation, it nevertheless is like-
wise the result of researches undertaken by the author with more
or less assiduity for a number of years, and, as a whole, it offers
a considerable number of deviations from any classification

hitherto submitted. It therefore seems proper, especially in view of the fact that this article will have a circulation among many persons who are interested in the collection and study of shells, but who have never paid especial attention to the principles of classification involved in the arrangement of the mollusks, to offer a few prefatory remarks on Taxonomy, or the science of classification, especially so far as those animals are concerned, and to answer the questions that may arise as to why some combinations are made.

PRIMARY DIVISIONS.

The classes of Mollusks are by no means allied to each other in equal degree ; there are two series that differ very widely, and which have been regarded by many of the best naturalists as primary groups of the animal kingdom ; that is, sub-kingdoms or branches. The great majority of the representatives of each of such groups do indeed offer so many special characteristics, and so widely differ from those of the other series, that perhaps the arguments in favor of such a view may be more weighty than those for the opposite. But the members of one class (Tunicata) seem to be in some respects intermediate or at least to narrow the chasm that would otherwise exist between the two, although their affinities are not regarded as dubious by most.

It has been found, after due investigation, that the central nervous system offers in its modifications in the Mollusks, as in the Vertebrates, the best criteria of relationship, and on the number of ganglia have been based the division thereof into the two primary groups, MOLLUSCA VERA and MOLLUSCOIDEA ; in the former (*Mollusca vera*), there are three well developed pairs of ganglia—the cerebral, the pedal, and the so-called branchial (or parieto-splanchnic of Huxley)—each pair being united by commissures ; in the latter (*Molluscoidea*), there is but one well developed pair, homologous with the pedal ganglia of the true Mollusks. Prof. Huxley, that very able biologist who has so much contributed by his clear mind and convincing logic to the education of the younger naturalists of the present day, has well remarked on the impossibility, or at least difficulty not yet surmounted, of the enunciation of a diagnosis which will combine the two divisions, and distinguish that combination from others.

And that difficulty has been strikingly illustrated by the positive withdrawal, by an able naturalist, of at least the Brachiopods and Bryozoans from the true Mollusks, and the combination of them with the Worms. If, then, a deviation from the example of Prof. Huxley and other masters in systematic zoology has been ventured in still retaining the combination of the two groups under the common branch name of Mollusca, it has been because there is still a certain conventional convenience in so doing, and because some members of the lower group (the Brachiopods) are almost always—at least by collectors—considered in connection with the higher forms. Another and more scientific reason is that at the confines of the lower groups, the hiatus between the two appears disproportionately little compared to that between the other branches, and a stricter series of homologies are traceable between the two. *Rhodosoma* (*Schizascus*, St.) of the Tunicates, and the recently described *Rhabdopleura*, Allmann, of the Bryozoans, are especially noticeable in this connection. It may also be added that the difficulty of framing a common diagnosis for the combined types appears to be the result of the diversity of secondary modifications and ramifications, and the extreme specialization of some forms and loss of common primitive characters, rather than of the divergence of the two types from a generalized *Proto-zoon* or aboriginal primordial stock—an element necessary to be considered in appreciation of the values of groups. In such cases, the test must be a series of consecutive inductions, and if those can be rigorously established, the truth cannot be far distant, even though an exclusive diagnosis cannot be applied. Care, however, must be taken not to abuse the privilege of combination without exact diagnosis, and the same latitude is not allowable in smaller and subordinate groups as in the more comprehensive.

CLASSES.

With regard to the classes of Mollusks, it is only necessary to state that the Pteropods have been considered as a subclass of Gasteropods, and thus retained in one and the same class with the typical members of the latter, in accordance with the views of most American malacologists, and because the hiatus between them appears to be much less than that between the Cephalopods and Gasteropods, and of course between those Odontophorous Mollusks and the Conchifers. The Pulmonifers of Cuvier—by some considered as a class apart—and the Solenoconchs—by some considered as also entitled to classic rank, by others referred to the Pteropods, and by others still to the Conchifers—have also been retained as sub-classes of the Gasteropods. The classification thus accepted is then the same as those already proposed, in 1861, by Prof. Dana[1] in his "Manual of Geology," and, in 1865, by Prof. E. S. Morse in his "Classification of the Mollusca based on the principle of cephalization." So far as the combination of the Pteropods, Heteropods, and typical Gasteropods into one class, others had also long before indicated the propriety of the innovation. The other groups regarded as of approximately equal value with those, and therefore designated sub-classes, are the Pectinibranchiates and Opisthobranchiates.

ORDERS.

Applying to the combinations of the Gasteropods into orders the principle that morphology and not teleology is the guide in natural classification, it becomes necessary to depart from some quite generally accepted schemes, and especially that whereby all the air-breathing mollusks are combined together in contradistinction from those respiring by means of branchiæ. As was perceived long ago by Cuvier, the inoperculated Pulmonifers (except *Proserpinidæ*) are entirely different from the operculated ones. That great naturalist very justly retained alone in one group the former (the *Proserpinidæ* were unknown to him), and thus constituted a truly natural order, while the operculated ones (*Cyclostomæ*, etc.) were referred to the Pectinibranchiates, and near *Littorina*, with which the best naturalists still associate them. His ignorance of the structure of the *Helicinidæ* induced him to retain them near the *Cyclostomæ*, but had he been acquainted with them, he would doubtless have combined them with his Trochoidea as they now are. The combination of all the Pulmoniferous Gasteropods into one group, as was afterwards done, was

[1] Prof. Dana has only differed in the depreciation of the value of the primary groups, the *Mollusca* (his ordinary *Mollusca*) and the *Molluscoidea* (his *Anthoid Mollusca*) being considered as classes, and their subdivisions as orders.

a decidedly retrograde step, and thus morphology was entirely subordinated to teleology, and even to a degree seldom equalled in recent times; for the groups enumerated are so very distinct from each other that they have no characters in common except those which they share with others as members of the same class, and the ability to breathe air direct—and even the adaptation for the latter office is affected by different modifications in the several subclasses.

The Heteropods, instead of representing a distinct class or subclass, are perhaps scarcely entitled to ordinal rank, but, as their distinctive characters are not entirely adaptive, they have for the present been accredited with it. Besides the *Dentalia* (So-LENOCONCHA), the *Chitonidæ* (POLYPLACOPHORA) have been removed from the association with the *Patellidæ* and *Acmaeidæ*, and for the last alone has been retained the ordinal name (DOCO-GLOSSA) proposed by Dr. Troschel for all the groups mentioned. It is difficult to understand why the *Chitonidæ* have been so persistently associated with *Patellidæ*, except for the reason that after the first discovery of the homologies between the two types, the great differences between them were in a measure lost sight of—a fault common to discoverers of unexpected relationships— and that most others have since been content to accept without active thought the approximation at first suggested. The similarity of the nervous system, recently urged in justification, seems to be more superficial than real, and rather the result of adaptation to the oval depressed form common to both. Although the author has been the first to limit (in manuscript long ago prepared) the order to the families now retained in it, the ordinal name proposed by Dr. Troschel (DOCOGLOSSA) being a suggestive one, it has been preferred to a new name.[1]

It need only be added that the orders of Conchifers and of all the Molluscoids are adopted simply as appearing to be the best that have been devised, and not because they are those likely to be ultimately confirmed, at least with precisely their present limits.

[1] Mr. W. H. Dall, after an extensive study of the anatomy of members of the group, had also arrived at the same conclusions, and was the first to demonstrate the entire want of affinity therewith of the Gadiniidæ.

FAMILIES.

The author has applied the views of those who consider those groups, above the rank of genera, combined by numerous common characters, and distinguished from neighboring groups by greater or more abrupt differences than those existing within the limits of such common associations, to be entitled to family rank. In Articulates, Vertebrates, and Radiates, such groups are often recognizable externally by a similarity of form which is dependent on more or less decided modifications of structure, or the relations between different parts. Very often, however—and especially in the Batrachians—such indications fail, and in the Mollusks there are many families that do not differ from each other in form; and, on the other hand, others exhibit a very considerable difference of form among their own representatives. Accepting the views as to the application of the term family to groups as adopted by the students of Mammals, we must apply them as we best can to the Mollusks, and of course we must be prepared for considerable diversity of views in the application, dependent on the personality of the observer, his acquaintance with the groups, and the path by which he has approached the study.

Very many, and probably most of the families now adopted, require revision based on more extensive materials than have yet been available to any one investigator. If any are to be especially pointed out in this connection, those of the orders of Cephalopods, and among the Gastropods, the *Turbinellidæ*,[1] *Pupinidæ*, and the sub-divisions of the disintegrated *Helicidæ*, *Melaniidæ*, *Cerithacea*, and *Trochacea*, may be indicated. But, because their affinities are doubtful, they have been for the present retained, for it is believed that the evils resulting from heterogeneous combinations (not definable by diagnosis) is greater than those resulting from refinement of analysis.

The acquaintance of the author with the Polyzoa being ex-

---

[1] The *Turbinellidæ* are retained as distinct on the authority of a very distinguished naturalist, who has kindly informed me that they are "*Stromboidæ*." I have not ventured to separate them, however, farther from the *Cynodontidæ* till more is known.

tremely limited, he has adopted without modification the classification of Bronn (who has availed himself of all the information
published up to his time), except for the Phylactolæmata, for
which he has followed Prof. Hyatt, who has since thoroughly
studied that order.

The details of classification of the families are yet too unsettled
to warrant the retention of the many sub-families which have
been proposed, and while the necessity for the adoption of such
subordinate groups is readily foreseen and admitted, so few
have been characterized in a manner which could be maintained
against criticism or justified by valid arguments, that only in
exceptional cases have any been admitted.

GENERAL CONSIDERATIONS.

In this connection it may be remarked that there is no scientific
basis for an *a priori* assumption that because the modifications
of an organ are of a certain importance in one branch or class
of animals, they are so in others. While such hints may perhaps
be of some use, the value requires to be *verified* in each instance.
Because the modifications in structure of the heart in mammals,
birds, and reptiles are of prime importance, it does not follow
that they are equally so in batrachians and fishes, and such a view
is, indeed, opposed to facts. Still less foundation exists for the
a priori application of such ideas to the classification of the
mollusks; and their distribution into two series, distinguished by
the bilocular (Monotocardian) and trilocular (Diocordian) par-
tition of the heart, certainly seems to be opposed by the indications furnished by the sum total of the organization.

And in like manner, because the modifications of a certain part
are the best indexes of affinity in one group of a class, it does
not follow that even in the same class, in another group, analogous modifications are of like value. The dentition, for example,
is quite characteristic in the mammalian orders Carnivores,
Ungulates, and Rodents; but in the Implacentals the value of
analogous modifications is very much less, and, within the range
of the same order (Marsupials), superficial differences, apparently
at least, as great as those between the cited orders of Placen-
tals are found. If, therefore, the modifications of the dentition
are used for the distinction of orders in one case, it is not because

the dentition is the most important *per se*, but because, as a matter of fact and experience, it has been determined that the modifications thereof are the co-ordinates of corresponding, though perhaps not as readily recognizable, modifications of other parts, and being so, they are taken advantage of for diagnostic purposes.

In like manner, as a matter of experience, the groups of the Pectinibranchiate mollusks agreeing in the dentition of the radula appear to agree in other important respects, and therefore the modifications of the teeth of the radula have been made use of as the prime characters, *because* they appear to be the exponents of the sum total of structure, and until it is *shown*, by a study and co-ordination of the modifications of the entire structure, that there are other characteristics that are of more importance and better indexes of affinity, and the application has been actually made, it is not evident what other better combinations capable of demonstration and diagnosis—the true criteria—can be made. Undoubtedly we have much yet to learn concerning the affinities of all the mollusks, and undoubtedly very considerable, and perhaps fundamental, modifications of classification will be required; but, in addition to objections against a given system, suggestions for reform are at the same time desirable, and then a comparison of the respective merits of the competing systems can be instituted.

As it is evident that the differences of dentition in the Placental and Implacental mammals is of very unequal value, it is no more than might be expected that the dentition in the class of Gasteropods should also vary in value, and it is actually found that while in the Pectinibranchiates the dentition is an excellent index of affinities, it is not so in the Tectibranchiates or Nudibranchiates. In this admitted fact, however, there is no more valid argument against its value in the Pectinibranchiates than in the corresponding case in mammals.

EXTINCT FORMS.

With respect to the extinct forms, the compiler has deemed it advisable to accept the views of the most approved students of the groups as to their relations, but has felt obliged to apply to them that indefinite but generally appreciated standard of value which has been used for the living forms, and consequently the number of extinct families admitted is larger than is generally recognized, especially in the class of Cephalopods. The views of M. Barrande have been implicitly accepted in the arrangement of the families of Tetrabranchiates, save as to the value of the groups. M. Barrande has designated the Mollusca as a *class*, the Cephalopoda as an *order* of that class, and has subdivided the latter into three *families*, each comprising a greater or less number of *genera*. The standard of value applied by that learned naturalist is in each case, but especially in the appreciation of the major groups, very different from that almost universally current, and as the more comprehensive groups are here retained with the higher rank generally accredited to them, the genera are also raised to a more elevated rank : the views of M. Barrande concerning the range of his genera being provisionally accepted, they are each one raised to family rank, and although the author is disposed to dissent from the positions assumed by M. Barrande in respect to the affinities and extent or relative value of certain of his genera, his knowledge of those forms is so vastly inferior to that naturalist's, that he has not ventured in any case to depart from him, even when he would have simply accepted the views of others, for none have had such opportunities for study, or made such good use of them, as he. As the expediency of the extension of family rank to some of the forms may be questioned, it may be remarked that the tendency of some naturalists seems to be to even subdivide still more minutely, Prof. Agassiz and Prof. Hyatt, for example, differentiating the genus *Ammonites* of most authors into a number of *families*, and separating ordinally the "Ammonoids" from the *Nautilidæ*.

In addition to the numerous extinct types of the Cephalopods, there are undoubtedly many among the Gasteropods and Conchifers that are entitled to family rank; but in view of the inability of the author to study many of them, and of our ignorance of their relations, it has been deemed inadvisable to name them.

SYNONYMY.

In order to make known the extent of the families adopted, as well as to direct students to reliable sources of information, reference has been made to a specific authority for each family.

It has been deemed preferable, however, all other things being equal, to refer to some readily accessible and popular work. But in cases where such works do not give the limits to the families which have been indicated by the most approved researches, references are made to the monographs or other publications wherein the information is furnished. Some of the families, however, have not yet been assigned the limits which, in the opinion of the compiler, appear the most natural; in order, therefore, to indicate as nearly as possible the relative values of the respective groups, the system of notation recommended especially by the late Hugh Strickland has been adopted. When there is an exact equivalency, either as to the limits assigned by the diagnosis, or as to the contents, the sign of equality ($=$) is used; when the group referred to is larger than that adopted, the corresponding sign ($<$) is prefixed to the former; when the group referred to is smaller, the usual sign indicative thereof ($>$) is prefixed; and when the group referred to is entirely different, including some forms not in and excluding others retained in the group compared with it, the sign ($\times$) is employed as a prefix.

ACKNOWLEDGMENTS.

In the appended list of authorities, and in connection with the names of the families, will be found the references to those authors who have been followed in especial cases. The compiler would also especially acknowledge his obligations to Mr. W. H. Dall for various kind offices and assistance in the preparation of this list.

ARRANGEMENT

OF

FAMILIES OF MOLLUSKS.

[Adopted provisionally by the Smithsonian Institution.]

N. B.—The Fossil Families are in Italics.

Class A.—CEPHALOPODA.

Order I.—DIBRANCHIATA.

Sub-Order Octopoda.

(*O. littorales.*)

1. Cirrhoteuthidae   < Octopodidae, Ad. I, 18.
2. Octopodidae   < Octopodidae, Ad. I, 18.

(*O. pelagici.*)

3. Philonexidae   = Philonexidae, Ad. I, 21.
4. Argonautidae   = Argonautidae, Ad. I, 23.

Sub-Order Sepiophora.

(*Oigopsidae.*)

5. Cranchiidae   { Cranchiidae, Ad. I, 26.
               { Loligopsidae, Ad. I, 27.
6. Chiroteuthidae   = Chiroteuthidae, Ad. I, 28.
7. Onychoteuthidae   < Onychoteuthidae, Ad. I, 30.
8. Ommastrephidae   < Onychoteuthidae, Ad. I, 30.

(Myopsidae.)

9. Loliginidae   < Loliginidae, Ad. I, 35.
10. Sepiolidae   < Loliginidae, Ad. I, 41.
11. Sepiidae   = Sepiidae, Ad. I, 41.
12. *Belosepiidae*   < Sepiidae, Chenu I, 46.
13. Spirulidae   = Spirulidae, Ad. I, 44.
14. *Belopteridae*   < Spirulidae, Chenu I, 51.
15. *Belemnitidae*   = *Belemnitidae*, Chenu I, 46.

ORDER II.—TETRABRANCHIATA.

(Nautiloidea.)

*
16. *Nothoceratidae*   = *Nothoceras*, Barr. II, 72.
17. *Bathmoceratidae*   = *Bathmoceras*, Barr. II, 74.

*
18. *Trochoceratidae*   = *Trochoceras*, Barr. II, 74.

*
19. Nautilidae   = Nautilus, Barr. II, 128.
20. *Hercoceratidae*   = *Hercoceras*, Barr. II, 152.
21. *Gyroceratidae*   = *Gyroceras*, Barr. II, 156.
22. *Lituitidae*   = *Lituites*, Barr. II, 168.
23. *Phragmoceratidae* = *Phragmoceras*, B. II, 189.
24. *Gomphoceratidae*   = *Gomphoceras*, B. II, 243.
25. *Cyrtoceratidae*   = *Cyrtoceras*, Chenu I, 73.
26. *Orthoceratidae*   > *Orthoceras*, Chenu I, 59.

*
27. *Ascoceratidae*   { *Ascoceras*, Barr. II, 334.
                  Aphragmites, Barr. II, 366.
                  Glossoceras, Barr. II, 372.

(*Goniatitoidea.*)

28. *Clymeniidae*	=	*Clymeniidae*, Chenu I, 70.
29. *Goniatitidae*	=	*Goniatites*, Chenu I, 75.
30. *Bactritidae*	=	*Bactrites*, Chenu I, 77.

(*Ammonitoidea.*)

31. *Turrilitidae*
$\left\{ \begin{array}{l} \textit{Turrilites, Chenu I, 95.} \\ \textit{Helicoceras, Chenu I, 96.} \\ \textit{Heteroceras, Chenu I, 96.} \end{array} \right.$

*

32. *Ceratitidae*	=	*Ceratites*, Chenu I, 76.
33. *Ammonitidae*	=	*Ammonites*, Chenu I, 77.
34. *Scaphitidae*	=	*Scaphites*, Chenu I, 91.
35. *Crioceratidae*	=	*Crioceras*, Chenu I, 90.
36. *Ancyloceratidae*	=	*Ancyloceras*, Chenu I, 92.
37. *Hamitidae*	=	*Hamites*, Chenu I, 93.
38. *Ptychoceratidae*	=	*Ptychoceras*, Chenu I, 94.
39. *Hamulinidae*	=	*Hamulina*, Chenu I, 94.
40. *Toxoceratidae*	=	*Toxoceras*, Chenu I, 93.

*

41. *Baculitidae*	=	*Baculites*, Chenu I, 95.
42. *Baculinidae*	=	*Baculina*, Chenu I, 77.

Class B.—GASTEROPODA.

Sub-Class Diœca.

Order III.—PECTINIBRANCHIATA.

Sub-Order Toxoglossa.

43. Conidae	= Conoidea, Tr. 16.
44. Pleurotomidae	= Pleurotomacea, Tr. II, 38.
45. Melatomidae	= Clionellidae, Stm. A. J. C. 1865, 62.
46. Haliidae	= Haliacea, Tr. II, 36.
47. Terebridae	= Terebracea, Tr. II, 27.
48. Cancellariidae	= Cancellariacea, Tr. II, 45.
49. Admetidae	= Admetacea, Tr. II, 46.

Sub-Order Rhachiglossa.

(*Typica.*)

50. Cystiscidae	= Cystiscidae, Stm. A. J. C. 1865, 55.
51. Marginellidae	< Marginellacea, Tr. II, 57.
52. Volutidae	= Volutacea, Tr. II, 54.
a. Volutomitrinae	{ Volutomitrina, Gray, 36. Amoriana, Gray, 35.
b. Volutinae	{ Volutina, Gray, 32. Yetina, Gray, 32.

5

(*Odontoglossa.*)

53. Fasciolariidae	= Fasciolariacea, Tr. II, 60.
a. Fusinae	
b. Fasciolariinae	
54. Mitridae	= Mitracea, Tr. II, 66.

(*Duplohamata.*)

55. Melongenidae	= Cassidulina, Tr. II, 79.
56. Buccinidae	< Fusacea, Tr. II, 69.
a. Photinae	= Photina, Tr. II, 82.
b. Buccininae	= Buccinina, Tr. II, 69.
c. Chrysodominae	= Neptunina, Tr. II, 72.
57. Nassidae	= Nassacea, Tr. II, 87.
a. Cyclonassinae	
b. Nassininae	
58. Cynodontidae	< Fusacea, Tr. II, 69.
a. Cynodontinae	= Vasina, Tr. II, 84.
b. Imbricariinae	= Imbricariina, Tr. II, 86.
?59. Turbinellidae	< Vasidae, Ad. I, 155.

(*Hamiglossa.*)

60. Turridae	= Strigatellacea, Tr. II, 202.
61. Olividae	= Olivacea, Tr. II, 105.
a. Olivinae	= Dactylina, Tr. II, 107.
b. Olivellinae	= Olivellina, Tr. II, 110.
c. Ancillinae	= Ancillina, Tr. II, 111.
62. Harpidae	= Harpacea, Tr. II, 104.

63. Ptychatractidae = Ptychatractidae, Stm. A. J. C. 1865, 59.
64. Muricidae
   a. Muricinae = Muricea, Tr. II, 112.
   b. Purpurinae = Purpuracea, Tr. II, 124.

(*Atypoglossa.*)
65. Columbellidae = Columbellacea, Tr. II, 97.

SUB-ORDER TÆNIOGLOSSA.
GROUP ROSTRIFERA.

66. Pomatiidae = Pomatiacea, Tr. I, 65.
67. Cyclostomidae = Cyclostomacea, Tr. I, 68.
   a. Licincinae = Licinea, Pfr. Pneum.
   b. Cistulinae = Cistulea, Pfr. Pneum.
   c. Cyclostominae = Cyclostomea, Pfr. Pneum.
68. Cyclophoridae = Cyclotacea, Tr. I, 66.
   a. Cyclotinae = Cyclotea, Pfr. Pneum.
   b. Cyclophorinae = Cyclophorea, Pfr. Pneum.
69. Pupinidae
   a. Pupininae = Pupinea, Pfr. Pneum.
   b. Diplommatininae Diplommatinacea, Pfr. Pneum.

*

70. Aciculidae = Aciculacea, Tr. I, 65.
71. Truncatellidae = Truncatellacea, Tr. I, 85.

*

72. Ampullariidae = Ampullariacea, Tr. I, 86.

*

73. Valvatidae = Valvatae, Tr. I, 95.

74. Viviparidae = Viviparidae, Gill. P. A. N. S. P. 1863, 33.
a. Lioplacinae = Lioplaces, Gill, P. A. P. '63.
b. Viviparinae = Vivipari, Gill, P. A. P. '63.

75. Assiminiidae < Assiminiidae, Ad. II, 314.
76. Rissoellidae = Rissoellidae, Ad. I, 325.
77. Pomatiopsidae = Pomatiopsinae, Stm. Hydr. 4, 29–36.
78. Rissoidae < Rissoidae, Stm. Hydr. 3.
a. Amnicolinae = Hydrobiinae, Stm. Hydr. 5.
b. Rissoinae = Rissoinae, Stm. Hydr. 5.
c. Rissoininae = Rissoininae, Stm. Hydr. 5.
79. Skeneidae = Skeneinae, Stm. Hydr. 5.
80. Bythiniidae = Bythiniinae, Stm. Hydr. 5.
81. Fossaridae = Fossari, Tr. I, 153.
82. Littorinidae > Littorinae, Tr. I, 129.
a. Lacuninae
b. Littorininae ?

83. Pyramidellidae = Pyramidellidae, Ad. I, 228.
84. Eulimidae = Eulimidae, Ad. I, 235.
85. Styliferidae = Styliferidae, Ad. I, 238.

86. Ceriphasiidae = Strepomatidae, Tr'n A. J. C. 1865.
87. Melanopidae = Pachycheili, Tr. I, 113.
88. Melaniidae

a. Melaniinae   = Melaniae, Tr. I, 121.
b. Tiarinae   = Thiarae, Tr. I, 112.
c. Paludominae
89. Cerithiopsidae   < Cerithia, Tr. I, 139.
90. Cerithiidae   < Cerithiacea, Tr. I, 138.
   a. Cerithiinae   < Cerithia, Tr. I, 139.
   b. Potamidinae   = Potamides, Tr. I, 145.
91. Planaxidae   < Planaxes, Tr. I, 149.
92. Caecidae   = Caecidae, Cpr. P. Z. S. 1858, 413.
93. Vermetidae   < Vermetacea, Mch.P.Z.S.1861, 1862.
94. Tenagodidae   < Vermetacea, Mch. P. Z. S. 1861, 1862.
95. Turritellidae   = Turritellae, Tr. I, 152.
   *
96. Trichotropidae   = Trichotropidae, Tr. I, 164.
   *
97. Hipponicidae   = Hipponicidae, Tr. I, 162.
98. Capulidae   < Capulacea, Tr. I, 156.
99. Calyptriidae   = Calyptræidae,Gray,P.Z.S.'67. 726.
   *
100. Neritopsidae   = Neritopsidae, Gray 51.
   *
101. Onustidae   = Onustidae, Tr. I, 190.
   *
102. Strombidae   = Alata, Tr. I, 191.
   a. Strombinae   = Strombinae, Gill, A. J. C. 1870

b. Seraphyinae   = Seraphyinae, Gill, A. J. C. 1870.

*

103. Aporrhaidae   = Aporrhaidae, Tr. I, 199.

(*Digitiglossa.*)

104. Pediculariidae = Pediculariacea, Tr. I, 189.
105. Amphiperasidae = Amphiperasidae, Tr. I, 216.

ROSTRUM WITH INVERTIBLE TIP.

106. Cypraeidae   = Cypraeacea, Tr. I, 201.
    a. Cypraeinae
    b. Pustulariinae
107. Triviidae   = Triviacea, Tr. I, 214.
    a. Triviinae
    b. Eratoinae

*

108. Marseniidae   = Marseniidae, Tr. I, 185.
109. Velutinidae   = Velutinidae, Tr. I, 165.
110. Naticidae   = Naticacea, Tr. I, 169.

GROUP PROBOSCIDIFERA.

111. Pyrulidae   = Sycotypidae, Tr. I, 238.
112. Doliidae   = Doliacea, Tr. I, 224.
113. Cassididae   = Cassidea, Tr. I, 220.
114. Ranellidae   = Ranellacea, Tr. I, 227.
115. Tritonidae   = Tritoniacea, Tr. I, 231.

116. Ianthinidae = Ianthinidae, Gray, Guide, 53.
117. Solariidae = Architectonidae, Gray, Guide, 62.
118. Scalariidae = Scalariadae, Gray, Guide, 52.

ORDER IV.—HETEROPODA.

119. Atlantidae = Atlantacea, Tr. I, 41.
120. Carinariidae = Carinariacea, Tr. I, 42.
121. Pterotrachaei- = Firolacca, Tr. I, 43.
dae

ORDER V.—RHIPHIDOGLOSSA.

SUB-ORDER PODOPHTHALMA.

(*Pseudobranchia.*)

122. Hydrocaenidae = Hydrocaenacea, Tr. I, 83.
123. Stoastomidae = Stoastomidae, Chitty, P. Z. S. 1857, 162.
124. Helicinidae = Helicinacea, Tr. I, 75.
125. Proserpinidae = Proserpinacea, Tr. I, 84.

(*Neritacea.*)

126. Neritidae = Neritinidae, Gray, 136.

(*Trochacea.*)

127. Rotellidae = Rotelladae, Gray, 139.
128. Turbinidae = Turbinidae, Gray, 141.
129. Liotiidae = Liotiadae, Gray, 146.

130. Trochidae   = Trochidae, Gray, 147.
131. Stomatellidae   = Stomatellidae, Gray, 158.

(*Pleurotomariacea?*)

132. Pleurotomarii-   < Pleurotomaridae, Br. Kef. Th.
      dae   III, 1037.
133. Scissurellidae   = Scissurellidae, Gray, 160.

(*Haliotacea.*)

134. Haliotidae   = Haliotidae, Gray, 161.
                         ?

(*Macluraeacea.*)

135. Macluraeidae   = Maclureadae, Cpr., Lect. 68.

SUB-ORDER DICRANOBRANCHIA.

(*Fissurellacea.*)

136. Fissurellidae   < Fissurellidae, Gray, 162.
137. Emarginulidae   < Fissurellidae, Gray, 162.
                         ?

(*Bellerophontacea.*)

138. Bellerophontidae = Bellerophontidae, Meek, P. C.
                     A. S., I, 9.

ORDER VI.—DOCOGLOSSA.

SUB-ORDER PROTEOBRANCHIA.

139. Acmaeidae   = Acmaeidae, Dall, A. J. C. 1870.
140. Patellidae   = Patellidae, Dall, A. J. C. 1870.

SUB-ORDER ABRANCHIA.

141. Lepetidae   = Lepetidae, Dall, A. J. C. 1869, 140.

ORDER VII.—POLYPLACOPHORA.

142. Chitonidae   < Chitonidae, Gray, 177.
143. Chitonellidae   < Chitonidae, Gray, 177.

SUB-CLASS PULMONIFERA.

ORDER VIII.—PULMONATA.

SUB-ORDER GEOPHILA.

(*Oculiferous tentacles invertible.*)

(*Agnatha.*)

144. Oleacinidae   < Testacellea, Alb. Mart. 22.
145. Streptaxidae   = Streptaxidae, Gray, A.M.N.H. VI, 1860, 268.
146. Testacellidae   < Testacellea, Alb. Mart. 22.

(*Goniognatha.*)

147. Orthalicidae   = Orthalicea, Alb. Mart. 209.

(*Holognatha.*)

148. Cylindrellidae = Cylindrellidae, Cr. & F., J. C. 1870, 5.
149. Pupidae   < Pupacea, Alb. Mart. 228.

150. Helicidae   < Helicacea, Alb. Mart. 80.
151. Vitrinidae   = Vitrinea, Alb. Mart. 43.

(*Togata.*)

152. Philomycidae   = Philomycenidae,Gray,A.M.N. H. VI, 1860, 269.

(*Subnuda.*)

153. Cryptellidae   = Cryptellidae,Gray, A.M.N.H. VI, 1860, 269.
154. Parmacellidae   = Parmacellidae, Gray, A.M.N. H. VI, 1860, 268.

*

155. Limacidae   < Limacidae, Ad. II, 217.
156. Arionidae   = Arionidae, Ad. II, 227.

(*Elasmognatha.*)

157. Succinidae   = Succinea, Alb. Mart. 308.
158. Janellidae   = Janellidae, Ad. II, 227.

(*Oculiferous tentacles simply contractile.*)

159. Vaginulidae   = Veronicellidae, Ad. II, 231.
160. Onchidiidae   = Onchidiidae, Ad. II, 232.

SUB-ORDER BASOMMATOPHORA.

(*Limnophila.*)

161. Chilinidae   = Chilinidae, Dall, A. L. N. Y. IX, 357, 1870.
162. Physidae   = Physidae, Dall, A. L. N. Y. IX, 355, 1870.

163. Ancylidae = Ancylidae, Dall, A. L. N. Y. IX, 354, 1870.

164. Limnaeidae = Limnaeidae, Dall, A..L. N. Y. IX, 348, 1870.

165. Otinidae = Otininae, Ad. I, 249.

166. Auriculidae = Ellobiinae, Ad. I, 236.

(*Petrophila.*)

167. Siphonariidae = Siphonariidae, Dall, A. J. C. 1870, 8.

168. Gadiniidae = Gadiniidae, Dall, A. J. C. 1870, 30.

(*Thalassophila.*)

169. Amphibolidae = Amphibolidae, Ad. II, 268.

SUB-CLASS OPISTHOBRANCHIATA.

ORDER IX.—TECTIBRANCHIATA.

A

170. Philinidae < Philinidae, Gray, 191.

171. Amphyspiridae = Amphyspiradae, Gray, 194.

172. Ringiculidae = Ringiculidae, Meek, C. L. I. F. N. A., Cret., 16, 34.

173. Actaeonidae < Actaeonidae, Meek, Sill. J. XXXV, 84.

174. *Actaeonellidae*   < *Actaeonidae*, Meek, Sill. J.
                    XXXV, 84.
                    *

175. Cylichnidae   = Bullinadae, Gray, 195.
                    *

176. Bullidae   = Bullidae, Gray, 196.

177. Amplustridae   = Amplustridae, Gray, 197.
                    *

178. Lophocercidae   = Lophocercidae, Gray, 201.

179. Aplysiidae   = Aplysiadae, Gray, 198.

                    B.

180. Runcinidae   = Runcinadae, Gray, 204.
                    *

181. Tylodinidae   = Tylodinadae, Gray, 203.

182. Umbrellidae   = Umbrelladae, Gray, 204.

183. Pleurobranchii-
      dae   = Pleurobranchidae, Gray, 201.

ORDER X.—NUDIBRANCHIATA.

SUB-ORDER PYGOBRANCHIA.

184. Doridopsidae   = Doridopsidae, A. & H., T. Z. S.
                    1864, 124.
                    *

185. Dorididae   = Dorididae, Gray, 208.

186. Onchidorididae = Onchidoridae, Gray, 206.
                    *

187. Goniodorididae = Goniodoridae, Gray, 211.

188. Polyceridae   < Polyceradae, Gray, 213.

189. Triopidae   > Triopidae, Gray, 214.

190. Ceratosomidae = Ceratosomidae, Gray, 215.

SUB-ORDER POLYBRANCHIA.

(*Inferobranchia.*)

191. Phyllidiidae = Phyllidiadae, Gray, 216.
192. Diphyllidiidae = Diphyllidiadae, Gray, 216.

(*Polybranchia.*)

193. Tritoniidae = Tritoniadae, Gray, 217.
194. Scyllaeidae = Scyllaeidae, Gray, 218.

(*Ceratobranchia.*)
(*Section* 1.)
(*A.*)

195. Dendronotidae = Dendronotidae, Gray, 219.
196. Heroidae = Heroidae, Gray, 221.
197. Tethyidae = Tethyadae, Gray, 219.
198. Dotoidae = Dotonidae, Gray, 222.
199. Proctonotidae = Proctonotidae, Gray, 220.
200. Glaucidae = Glaucidae, Gray, 222.

(*B.*)

201. Eolididae = Eolididae, Gray, 223.

(*Section* 2.)

202. Fionidae = Fionidae, Gray, 227.
203. Hermaeidae = Hermaeidae, Gray, 227.

SUB-ORDER PELLIBRANCHIATA.
(*Tribe* 1.)

204. Elysiidae = Elysiadae, Gray, 228.
205. Limapontiidae = Limapontiadae, Gray, 229.

(*Tribe* 2.)

206. Phyllirrhoidae = Phyllirhoidae, Gray, 230.
?

SUB-ORDER ENTOCONCHACEA.

207. Entoconchidae = Heterosalpinx, Baur, N. A. A.
L. C. XXXI.

SUB-CLASS PTEROPODA.

ORDER XI.—THECOSOMATA.

208. Limacinidae = Limacinacea, Tr. I, 50.
209. Hyalidae = Hyalacea, Tr. I, 50.
210. Cymbuliidae = Cymbuliacea, Tr. I, 53.
211. *Conulariidae* = *Conulariidae*, Br. Th. III, 645.
212. *Hyolithidae* = *Thecidae*, Br. Th. III, 646.

ORDER XII.—GYMNOSOMATA

214. Clionidae = Clionacea, Tr. I, 54.
*

215. Pneumodermo-
nidae = Pneumodermacea, Tr. I, 56.
*

216. Cymodocceidae = Pterocymodoccidae, Br. Th.
III, 645.

SUB-CLASS PROSOPOCEPHALA.

ORDER XIII.—SOLENOCONCHÆ.

217. Dentaliidae = Dentaliidae, Br. Th. III, 523.
2

Class C.—CONCHIFERA.

Order XIV.—DIMYARIA.

(Pholadacea.)

218. Aspergillidae	< Gastrochaenidae, Tryon, P. A. N. S. P., 1861, 465.
219. Gastrochaenidae	< Gastrochaenidae, Tryon, P. A. N. S. P., 1861, 465.
220. Teredinidae	= Teredidae, Tryon, P. A. N. S. P., 1862, 453.
221. Pholadidae	= Pholadidae, Tryon, P. A. N. S. P., 1862, 191.

(Solenacea.)

222. Solenidae	< Solénacées, Desh. 1860, 143.
223. Solecurtidae	< Solénacées, Desh. 1860, 143.

(Myacea.)

224. Saxicavidae	= Glycimérides, Desh. 1860, 165.
225. Myidae	< Myaires, Desh. 1860, 182.
226. Corbulidae	< Myaires, Desh. 1860, 182.
227. Pandoridae	= Pandoridae, Desh. 1860, 238.
228. Anatinidae	< Osteodesmidae, Desh. 1860, 245.
229. Myochamidae	= Myochamidae, Cpr. Lect. 103.

*

230. Pholadomyidae = Pholadomyadae, Desh. 1860, 270.

(*Veneracea.*)

231. Mactridae   < Mactracea, Desh. 1860, 281.
232. Mesodesmidae   = Mésodesmides, Desh. 1860, 297.
233. Amphidesmidae = Amphidesmidae, Desh. 1860, 297.

*

234. Tellinidae   = Tellinidae, Desh. 1860, 314.
235. Psammobiidae = Psammobidae, Desh. 1860, 364.
236. Donacidae   = Donacidae, Desh. 1860, 387.
237. Petricolidae   = Lithophaga, Desh. 1860, 400.
238. Veneridae   < Conchae, Desh. 1860, 407.
239. Glauconomidae = Glauconomyadae, Ad. II, 442.

(*Corbiculacea.*)

240. Cyrenidae   = Cycladae, Gray, Turton, 250.
241. Pisidiidae   = Pisidiidae, Gray, Turton, 263.
242. Cyrenoididae   = Cyrenoididae, Ad. II, 452.

(*Dreissenacea.*)

243. Dreissenidae   = Dreissenidae, Ad. II, 52.

(*Cardiacea.*)

244. Veniliidae   = Cyprinidae, Ad. II, 443.
245. Glossidae   < Bucardiidae Ad. II, 460.

246.	Cardiidae	< Cardiacca, Desh. 1860, 527.
247.	Adacnidae	< Cardiacca, Desh. 1860, 527.

(*Chamacea.*)

248.	Chamidae	= Chamacea, Desh. 1860, 577.

(*Lucinacea.*)

249.	Lucinidae	< Lucinidae, Desh. 1860, 588.
250.	Ungulinidae	< Ungulinidae, Ad. II, 470.
251.	Erycinidae	< Lascidae, Ad. II, 473.
252.	Cyamiidae	< Lascidae, Ad. II, 473.
253.	Leptonidae	< Leptonidae, Ad. II, 477.
254.	Galeommidae	< Galeommidae, Ad. II, 479.

(*Solemyacea.*)

255.	Solemyidae	= Solemyadae, Desh. 1860, 728.

(*Carditacea.*)

256.	Crassatellidae	= Crassatellidae, Desh. 1860, 733.
257.	Carditidae	= Carditae, Desh. 1860, 751.

(*Naiades.*)

258.	Unionidae	< Unionidae, Ad. II, 489.
259.	Iridinidae	= Mutelidae, Ad. II, 505.
260.	Mycetopodidae	= Mycetopodidae, Gray, P. Z. S., 1847, 197.

(*Muelleracea.*)

261.	Ætheriidae	< Ætheriidae, Ad. II, 509.
262.	Muelleriidae	< Ætheriidae, Ad. II, 509.

(*Trigoniacea.*)

263. Trigoniidae = Trigonea, Desh. 1860, 805.

(*Arcacea.*)

264. Nuculidae = Nuculidae, Ad. II, 544.
265. Ledidae = Ledidae, Ad. II, 546.
266. Arcidae = Arcacea, Desh. 1860, 832.

ORDER XV.—METARRHIPTAE.

267. Tridacnidae = Tridacnides, Vaill, A. S. N., IV, 1865, 64.

ORDER XVI.—HETEROMYARIA.

268. Mytilidae = Mytilidae, Ad. II, 511.

ORDER XVII.—MONOMYARIA.

(*Aviculacea.*)

269. Pinnidae = Pinnidae, Meek, Sill. J. XXXVII, 212.
270. Pteriidae = Pteriidae, Meek, Sill. J. XXXVII, 212.
271. Vulsellidae = Vulsellidae, Ad. II, 523.

(*Pectinacea.*)

272. Spondylidae = Spondylidae, Ad. II, 559.

273. Limidae = Radulidae, Ad. II, 556.
274. Pectinidae = Pectinidae, Ad. II, 550.

(*Anomiacea.*)

275. Placunidae = Placunidae, Carp. Lect. 123.
276. Anomiidae = Anomiadae, Carp. Lect. 123.

(*Ostracea.*)

277. Ostreidae = Ostracea, Ad. II, 567.

?

278. *Eligmidae* = *Eligmus*, Eudes Desl. M. L. S. N., X, 272.

?

Order XVIII.—RUDISTA.

279. *Hippuritidae* < *Hippuritidae*, Woodw. Man. 1866, 440.
280. *Radiolitidae* < *Hippuritidae*, Woodw. Man. 1866, 440.
281. *Caprinellidae* < *Hippuritidae*, Woodw. Man. 1866, 440.
282. *Caprinidae* < *Hippuritidae*, Woodw. Man. 1866, 440.
283. *Caprotinidae* < *Hippuritidae*, Woodw. Man. 1866, 440.

23

(Sub-Branch Molluscoidea.)

Class D.—TUNICATA.

Order XIX.—SACCOBRANCHIA.

(*Solitaria.*)

284. Pelonaeidae	= Pelonaeidae, Br. III, 216.
285. Chelyosomidae	< Ascidiadae, Br. III, 218.
286. Ascidiidae	< Ascidiadae, Br. III, 218.
287. Bolteniidae	< Ascidiadae, Br. III, 218.

*

287ᵃ· Rhodosomidae = Rhodosoma, Crosse, J. C. XV, 1877, 101.

(*Sociales.*)

(*S. Perophoracea.*)

288. Perophoridae   < Clavellinidae, Br. III, 217.

(*S. Clavellinacea.*)

289. Clavellinidae   < Clavellinidae, Br. III, 217.

(*Aggregata.*)

290. Sigillinidae	< Didemninae, Br. III, 217.
290ᵃ· Didemnidae	< Didemninae, Br. III, 217.
291. Leptoclinidae	< Didemninae, Br. III, 217.

*

292. Polyclinidae   < Polyclininae, Br. III, 217.
293. Synocciidae   < Polyclininae, Br. III, 217.
*
294. Botryllidae   = Botryllidae, Br. III, 217.

ORDER XX.—DACTYLOBRANCHIA.
295. Pyrosomidae   = Pyrosomatidae, Br. III, 216.

ORDER XXI.—TAENIOBRANCHIA.
296. Doliolidae   = Doliolidae, Br. III, 216.
*
297. Salpidae   = Salpidae, Br. III, 216.

ORDER XXII.—LARVALIA.
298. Appendicula-   = Appendiculariadae, Br. III,
riidae   216.

25

Class E.—BRACHIOPODA.

Order XXIII.—ARTHROPOMATA.

(*Ancylopoda.*)

299. Terebratulidae	< Terebratulidae, Dav. Int. 61.
a. Terebratuli-nae	= Terebratulinae, Dall, A. J. C. 1870.
b. *Stringocepha-linae*	= *Stringocephalinae*, Dall, A. J. C. 1870.
c. Magasinae	= Magasinae, Dall, A. J. C. 1870.
d. Kraussininae	= Kraussininae, Dall, A. J. C. 1870.
e. Platidiinae	= Platidiinae, Dall, A. J. C. 1870.
f. Megathyrinae	= Megathyrinae, Dall, A. J. 1870.
300. Thecidiidae	= Thecideidae, Dav. Int. 76.

(*Helictopoda.*)

301. *Spiriferidae*	< *Spiriferidae*, Dav. Int. 79.
302. *Atrypidae*	< *Spiriferidae*, Dav. Int. 90.
303. *Koninckinidae*	= *Koninckinidae*, Dav. Int. 92.
304. Rhynchonellidae	= Rhynchonellidae, Dav. Int. 93.
a. *Pentamerinae*	
b. Rhynchonelli-nae	

305. *Strophomenidae* = *Strophomenidae*, Dav. M. L. S. N., X, 191.

    a. Poramboniti-
      nae = *Porambonitidae*, Dav. Int. 99.

    b. Strophomeni-
      nae = *Strophomenidae*, Dav. Int. 101.

    c. Davidsoninae = *Davidsonidae*, Dav. Int. 109.

306. *Productidae* = *Productidae*, Dav. Int. 112.

Order XXIV.—LYOPOMATA.

307. Craniidae = Craniadae, Dav. Int. 123.

308. Discinidae = Discinidae, Dav. Int. 125.

309. Lingulidae = Lingulidae, Dall. A. J. C. VI, 1870.

    a. Lingulinae = Lingulinae, Dall. A. J. C. VI, 1870.

    b. Obolinae = *Obolinae*, Dall, A. J. C. VI, 1870.

Class F.—POLYZOA.

Order XXV.—PHYLACTOLÆMATA.

Sub-Order Lophopodia.

310. Pectinatellidae = Pectinatellidae, Hyatt, P. E. I. 1864-66.
311. Cristatellidae = Cristatellidae, Hyatt, P. E. I. 1864-66.
312. Plumatellidae = Plumatellidae, Hyatt, P. E. I. 1864-66.

Sub-Order Pedicellinea.

313. Pedicellinidae = Pedicellinidae, Bronn, III, 86.

Order XXVI.—GYMNOLÆMATA.

Sub-Order Urnatellea.

314. Urnatellidae = Urnatellidae, Bronn, III, 86.

Sub-Order Paludicellea.

315. Paludicellidae = Paludicellidae, Bronn, III, 86.

Sub-Order Chilostomata.

(*Incrustata* or *Rigida*.)

316. Selenariidae = Selenariadae, Bronn, III, 86.
317. *Steginoporidae* = *Steginoporidae*, Bronn, III, 86.

318. Eschariporidac = Eschariporidac, Bronn, III, 86.
319. Porcllinidac = Porcllinidac, Bronn, III, 86.
320. *Porcllidac* = *Porcllidac*, Bronn, III, 86.
321. Escharellidac = Escharellidac, Bronn, III, 86.
322. Escharellinidac = Escharellinidac, Bronn, III, 86.
323. Porinidac = Porinidac, Bronn, III, 86.
324. Escharinellidac = Escharinellidac, Bronn, III, 85.
325. Escharidac = Escharidac, Bronn, III, 85.
326. Flustrinidac = Flustrinidac, Bronn, III, 85.
327. Flustrellidac = Flustrellidac, Bronn, III, 85.
328. Flustrellariidac = Flustrellariadac, Bronn, III, 85.
329. Hippothoidac = Hippothoidac, Bronn, III, 84.

(*Radicellata.*)
(*Radicellata flexilia.*)

330. Gemellariidac = Gemellariadac, Bronn, III, 84.
331. Farciminariidac = Farciminariadac, Bronn, III, 84.
332. Flustridac = Flustridac, Bronn, III, 84.
333. Bicellariidac = Bicellariadac, Bronn, III, 84.
334. Electrinidac = Electrinidac, Bronn, III, 84.
335. Scrupariidac = Scrupariadac, Bronn, III, 83.

(*Radicellata articulata.*)

336. Salicornariidac = Salicornariadac, Bronn, III, 83.
337. Cellulariidac = Cellulariadac, Bronn, III, 83.
338. Catenicellidac = Catenicellidac, Bronn, III, 83.

29

SUB-ORDER CTENOSTOMATA.

339. Hislopiidae   = Hislopiadae, Bronn, III, 83.
340. Alcyonidiidae   = Alcyonidiadae, Bronn, III, 83.
341. Vesiculariidae   = Vesiculariadae, Bronn, III, 83.

SUB-ORDER CYCLOSTOMATA.
(*Articulata*.)

342. Crisiidae   = Crisiadae, Bronn, III, 82.

(*Inarticulata*.)
(*Inarticulata operculata*.)

343. Eleidae   = Eleidae, Bronn, III, 82.
344. Myriozoidae   = Myriozoidae, Bronn, III, 82.

(*Inarticulata fasciculata*.)

345. Fascigeridae   = Fascigeridae, Bronn, III, 82.
346. Fasciporidae   = Fasciporidae, Bronn, III, 81.

(*Inarticulata tubulata*.)

347. Tubigeridae   = Tubigeridae, Bronn, III, 81.
348. Sparsidae   = Sparsidae, Bronn, III, 80.
349. Clausidae   = Clausidae, Bronn, III, 80.
350. Crisinidae   = Crisinidae, Bronn, III, 80.
351. Cavcidae   = Cavcidae, Bronn, III, 79.

(*Inarticulata foraminata*.)

352. Ceidae   = Ceidae, Bronn, III, 79.
353. Cavidae   = Cavidae, Bronn, III, 79.
354. Cytidae   = Cytidae, Bronn, III, 79.
355. Crescidae   = Crescidae, Bronn, III, 79.

ORDER XXVII?—RHABDOPLEURAE.

356. Rhabdopleuri-   = Rhabdopleura, Allm. Q. J. M.
     dae   S., IX, n. s., 57.

LIST OF AUTHORS REFERRED TO.

The following enumeration of works is chiefly intended to explain the abbrevia-
tions used in connection with the preceding list of families, and as the works most
accessible to students generally have been used, whenever they could be referred
to in explanation of the limits of families adopted, titles of the most elaborate and
valuable monographs and catalogues of families and other groups have been entirely
omitted, although the compiler has been fortunate enough to be enabled to make
use of them. Special monographs have only been referred to when the groups in
connection with which they are cited have not been limited in the same manner in
general works.

In order, however, to facilitate the use of the list, as well as reference to the
series in question, Mr. Lovell Reeve's "Conchologia iconica" has been catalogued,
and all the monographs hitherto published enumerated, with references to the
families to which the respective genera belong in the present system.

For the information of students, and because it is information often desired, the
publishers' prices of most of the works cited are given, in the currency of the
country where they were published. Many of the separate monographs reprinted
from journals can be obtained from the second-hand book dealers—especially the
German—and from the Naturalists' Agency of Salem, Mass., but at varying prices.

In order to secure uniformity of typography, only the initial letters of the characteristic words are capital, the example of the learned brothers Grimm, as well as
other German writers, sanctioning such usage for their language. The punctuation
of the respective title-pages is adopted.

ADAMS (Henry *and* Arthur). The genera of recent Mollusca; arranged according to their organization. · · · · In three volumes. · · · · Vol. I. [-] III.
—— London : John Van Voorst, · · · · 1858. [8vo., V. I, 484 pp.; V. II, 2 p. l.
661 pp. ; Atlas, 3 p. l. 138 pl. w. 138 l. opposite. Published in 36 parts, 1st
Jan. 1853—1st Nov. 1858, at 2 sh. 6 d., plain ; 5 sh., animals colored, per part.]

ALBERS (Johann Christian). Die Heliceen nach natürlicher verwandtschaft
systematisch geordnet von Joh. Christ. Albers, · · · · Zweite ausgabe nach
dem hinterlassenen manuskript besorgt von Eduard von Martens. —— Leipzig, Verlag von Wilhelm Engelmann. 1860. [8vo., xviii. 359 pp.—3 th. 7½ ngr.]

ALDER (Joshua) *and* Albany HANCOCK. A monograph of the British
Nudibranchiate mollusca : with figures of all the species. · · · · London :
Printed for the Ray Society, 1845. [Imp. 4to., 5 p. l. 54 pp. 138 l., xl pp. 1 l.,
83 pl. Published in parts, 1845-55.]

[The arrangement of the Nudibranchiata is mostly adopted from Alder and
Hancock (op. cit. pp. xiv.—xxiv.). In place, however, of the single family

(31)

Dorididae, two (185, 186) are adopted; four (187, 188, 189, 190) instead of the Polyceridae, and two (195, 196) disintegrated from the Heroidae.]

ALDER (Joshua) *and* **Albany HANCOCK.** Notice of a collection of Nudibranchiate mollusca made in India by Walter Elliott, Esq., with descriptions of several new genera and species. (1863.) < Transactions of the Zoological Society of London, V, 1866, 113-147, pl. 28-33.

ALLMAN (George James). On Rhabdopleura, a new form of polyzoa, from deep-sea dredging in Shetland. < Quarterly Journal of Microscopical Science : [etc.], IX, n. s., 1869, 57—63, pl. 8.

AMERICAN Journal of Conchology. Volume I. [—] II. Edited by George W. Tryon, Jr., · · · · Philadelphia: George W. Tryon, Jr., 625 Market Street. 1865 [—] 1866. [Published quarterly, at $3 per number, or $10 per year.]
——The same. Volume III. [—] V. Published by the Conchological section of the Academy of Natural Sciences. · · · · Philadelphia : Conchological section of the Academy of Natural Sciences. · · · · 1867 [—] 1870. [Published at $10 per annum, payable in advance.]

BARRANDE (Joachim). Caractères distinctifs des Nautilides, Goniatides et Ammonides.—Établissement du genre Nothoceras, · · · · < Bulletin de la Société géologique de France. 2e série. XIII, 372-389, pl. 11—12, 1856.
[The genera enumerated in this article are co-equal with and arranged in the same sequence as the families of Goniatitoidea and Ammonitoidea, which are equivalent to the families Nautilides and Goniatides of Barrande.]
—— Système silurien du centre de la Bohême · · · · 1ère partie : Recherches paléontologiques. Vol. II. Texte. Classe des mollusques. Ordre des Céphalopodes. 1867. Chez l'auteur et éditeur | à Prague · · · · à Paris, · · · · [4to., xxxvi, 712 pp.—40 fr.]
—— The same. [Atlas.] 1ère partie: Recherches paléontologiques. Vol. II. Céphalopodes. [1ère—3me série, as below.] 1865 [-] 1868. Chez l'auteur et éditeur | à Prague, · · · · à Paris, · · · · [4to.]
1ère série : Planches 1 à 107. 1865. [100 fr.]
2me série : Planches 105 à 244. 1866. [125 fr.]
3me série : Planches 245 à 350. 1868. [140 fr.]

BAUR (Albert). Beiträge zur naturgeschichte der Synapta digitata. —— Dritte abhandlung : Die eingeweideschnecke (Helicosyrinx parasita) in der leibeshöhle der Synapta digitata. · · · · Dresden. Druck von E. Blochmann & sohn. 1864. [4to., 2 p. l. 119 pp. pl. vi.—viii.] < Novorum Actorum Academiæ Cæsareæ Leopoldino-Carolinæ naturæ curiosorum XXXI. 1864.

BRONN (Heinrich Georg). Die klassen und ordnungen des Thierreichs, wissenschaftlich dargestellt in wort und bild · · · · Dritter band. Malacozoa · · · · Leipzig und Heidelberg. C. F. Winter'sche verlagshandlung, 1862-66. [Published in 48 parts, 8vo., 1862-66, at ½ th. per part, and bound in 2 vols., with double titles, general and special, viz :]
III, 1. Die klassen und ordnungen der Weichthiere (Malacozoa), wissenschaftlich dargestellt in wort und bild. Von Dr. H. G. Bronn, · · · · Dritten band erste abtheilung. Kopflose Weichthiere (Malacozoa Acephala). · · · · [2 titles, pp. 1-518, pl. 44, w. 44 opp. expl. l. 1862.]

III, 2. Dr. H. G. Bronn's klassen und ordnungen der Weichthiere (Mala-cozoa), wissenschaftlich dargestellt in wort und bild. Fortgesetzt von Wilhelm Keferstein, M. D. . . . . Dritten bandes zweite abtheilung. Kopftragende Weichthiere (Malacozoa Cephalophora). . . . . [2 titles, pp. 521-1500, pl. 45-136, w. 92 opp. expl. 1. 1862-66.]

CARPENTER (Philip P . . .). First steps towards a monograph of the Cœ-cidæ, a family of rostriferous gasteropoda. < Proceedings of the Zoological Society of London. Part XXVI, 1858, 413—444.

—— Lectures on Mollusca; or, "shell-fish" and their allies. Prepared for the Smithsonian Institution, by Philip P. Carpenter, B. A., Ph. D., of Warrington, England. < Annual report of the board of regents of the Smithsonian Institution, . . . for . . . 1860, 1861, 151—283.
[Reprinted, with index, 140 pp., Washington, 1860.]

CHENU (Jean Charles). Manuel de conchyliologie et de paléontologie conchy-liologique par le Dr. J. C. Chenu . . . . Paris | Librairie Victor Masson . . . . 1859 [-] 62. [8vo., 2 v. I, 2 p. l. vii, 508 pp.; II, 3 p. l. 327 pp. Published in 3 parts, 1859-61 (@ 12.50 + 12.50 + 20 = 45 fr. ; reduced now to 32 fr.]

CHITTY (Edward). On Stoastomidæ as a family, and on seven proposed new genera, sixty-one new species, and two new varieties from Jamaica. < Pro-ceedings of the Zoological Society of London, Part XXV, 1857, pp. 162—201.

CROSSE (H . . .). Note sur un genre [Rhodosoma] intermédiare entre les ascidiens et les mollusques lamellibranches. < Journal de conchyliologie, v. XV (3e série, t. VII), 1867, 101—107.

CROSSE (H . . .) and Paul FISCHER. Étude sur la mâchoire et l'armature linguale des Cylindrellidæ et des quelques genres voisins sous la rapport con-chyliologique. < Journal de conchyliologie, v. XVIII (3o série. t. X), 1870, 5—27, pl. 3—5.

DALL (William Healey). Materials for a monograph of the family Lepetidæ. < American Journal of Conchology. V, 140—160, Pl. xv. 1870.

—— Materials toward a monograph of the Gadiniidæ. < Ib. VI, 8—22, pl. 2 and 4, fig. 1—3, 12—13. 1870.

—— Remarks on the anatomy of the genus Siphonaria, with a description of a new species. < Ib. VI, 30—41, pl. 4—5. 1870.

—— On the genus Pompholyx and its allies, with a revision of the Limnæidæ of authors. < Annals of the Lyceum of Natural History of New York. IX, 333—361; Pl. ii. 1870.

—— A revision of the Terebratulidæ and Lingulidæ, with remarks on, and de-scriptions of, some recent forms. < American Journal of Conchology. VI. 88—168, pl. 6, 7, and 8. 1870.

—— On the limpets; with special reference to the species of the west coast of America, and to a more natural classification of the group. < Ib. VI, 1870. (In press.)

DAVIDSON (Thomas). British fossil Brachiopoda. By Thomas Davidson, Esq., F.G.S., ... Vol. I. With a general introduction : I. On the anatomy of Terebratula. By Professor Owen, ... II. On the intimate structure of

3

the shells of the Brachiopoda. By Professor Carpenter, · · · · III. On the classification of the Brachiopoda. By Thomas Davidson, · · · · ── London: Printed for the Palæontographical Society. 1851—1854. [4to. 1 p. l. 136 pp. 9 pl. w. 9 l. expl.]

DAVIDSON (Thomas). Introduction à l'histoire naturelle des Brachiopodes vivants et fossiles, ou considérations générales sur la classification de ces êtres en familles et en genres; par Thomas Davidson, Esq., · · · · Traduit de l'Anglais par M. Eudes-Deslongchamps, · · · ; et par M. Eugene Eudes-Deslongchamps, · · · · <Mémoires de la Société linnéene de Normandie. X, 1856, 71—271, pl. 6—14, with 9 l. explan.

[A translation of the third part of the preceding work, with modifications by the author.]

DESHAYES (Gerard Paul). Description des animaux sans vertèbres découverts dans la bassin de Paris pour servir de supplément à la Description des coquilles fossiles des environs de Paris comprenant une revue générale de toutes les espèces actuellement connues, par G. P. Deshayes.—[See "Contents."]── Paris, J. B. Baillière et fils, · · · · 1860 [—] 1866. [50 livr., chaque livr. 5 fr.]

CONTENTS.

Tome premier.—Texte. Mollusques Acéphalés Dimyaires. Accompagné d'un Atlas de 89 planches. · · · 1860. [2 p. l. 912 pp.]

Tome deuxième.—Texte. Mollusques Acéphalés Monomyaires et Brachiopodes. Mollusques Céphalés. Première partie. Accompagné d'un Atlas de 64 planches. (Planches 1 à 64.) · · · 1864. [3 p. l. 968 pp.]

Tome troisième.—Texte. Mollusques Céphalés, deuxième partie. Mollusques Céphalopodes. Accompagné d'un Atlas de 42 planches. (Planches 65 à 107.) · · · 1866. [2 p. l. 667 pp.]

Atlas. Tome premier.—(89 planches.) Mollusques Acéphalés. · · · 1860. [2 p. l. [92] pp. [89] pl.]

Atlas. Tome deuxième.—(107 planches.) Mollusques Céphalés et Mollusques Céphalopodes. · · · 1866. [2 p. l. 107 pp. 107 pl.]

[This work is cited as containing the latest general revision of the classification of the Conchifera, by one who has perhaps devoted more attention to those animals than any other naturalist.]

DESLONGCHAMPS (Jacques Armand Eudes). Description d'un nouveau genre de coquilles bivalves fossiles Eligmus, provenant de la grande oolithe du département du Calvados ; · · · · <Mémoires de la Société linnéene de Normandie. X, 1856, 272—293, pl. 15—16.

GILL (Theodore Nicholas). Systematic arrangement of the mollusks of the family Viviparidæ, and others, inhabiting the United States. <Proceedings of the Academy of Natural Sciences of Philadelphia, 1863, 33—40.

── On the family Strombidæ, and its classification. <American Journal of Conchology. (Not yet published.)

GRAY (John Edward). Catalogue of the Mollusca in the collection of the British Museum. Part I. Cephalopoda Antepedia. Printed by order of the trustees. London: 1849. [12mo. viii, 164 pp.—4 sh.]

GRAY (John Edward.) A list of the genera of recent Mollusca, their synonyma and types. < Proceedings of the Zoological Society of London. Part XV, 1847, 129—219.

[Republished, with same pagination, and with special title-page, in "Figures of molluscous animals, selected from various authors. Etched for the use of students. By Maria Emma Gray." iv. 1859.]

—— On the arrangement of the Land Pulmoniferous mollusca into families. < The Annals and Magazine of Natural History. VI, Third Series, 1860, 267—269.

—— Notes on the specimens of Calyptræidæ in Mr. Cumming's collection. < Proceedings of the scientific meetings of the Zoological Society of London for the year 1867, 726—748.

HANCOCK (Albany). *See* Alder (Joshua) *and* Hancock.

HUXLEY (Thomas Henry). An introduction to the classification of animals. . . . . —— London : John Churchill & Sons, ... 1869. [8vo., 4 p. l. 147 pp. 6 sh.]

[Authority for the Tunicate order *Larvalia.*]

HYATT (Alpheus). Observations on Polyzoa. Suborder Phylactolæmata. < Proceedings of the Essex Institute, IV, V.

[Author's separate ed., iv, 103 pp., 15 pl. w. 7 intercalated leaves explanatory.]

JOURNAL de conchyliologie, comprenant l'étude des animaux, des coquilles vivantes et des coquilles fossiles, publié sous la direction de M. Petit de la Saussaye. Tome premier [—] quatrième. —— À Paris, chez M. Petit de la Saussaye, . . . , 1850 [—] 1853.

—— Journal de conchyliologie publié sous la direction de MM. Fischer et Bernardi. Tome V [—] VIII. 2e série. Tome Ier [—] IV. —— À Paris, chez M. Bernardi, ... Juillet 1856 [—] Janvier 1860.

—— Journal de conchyliologie, publié sous la direction de MM. Crosse et Fischer [et Bernardi, 1861—1863]. 8e série. Tome Ier [—] Xme. Vol. IX [—] XVIII. —— À Paris, chez M. Crosse, rue Tronchet, 25. 1861 [—] 1870.

[Prix de l'abonnement : pour France, 16 fr. ; pour les pays hors d'Europe, 20 fr.]

LEA (Isaac). A synopsis of the family Unionidæ. . . . Fourth edition, very greatly enlarged and improved. —— Philadelphia : Henry C. Lea. 1870. [4to. xxx pp. + bastard title + 25—184 pp.]

MACDONALD (John Denis). On the representative relationships of the fixed and free Tunicata, regarded as two subclasses of equivalent value ; with some general remarks on their morphology. Transactions of the Royal Society of Edinburgh. XXIII, 1864, 171—183, pl. ix, 1862-63.

—— On the anatomy and classification of the Heteropoda. < Ib. XXIII, 1864, 1—20, pl. i—ii, 1861-62.

MEEK (Fielding Bradford). Remarks on the family Actæonidæ, with descriptions of some new genera and subgenera. < The American journal of science and arts. Conducted by B. Silliman, B. Silliman, Jr., and James H. Dana [etc.]. Second series, XXXV, 1863, 84—94.

MEEK (Fielding Bradford). Remarks on the family Pteriidæ (= Aviculidæ) with descriptions of some new fossil genera. < American journal of science and arts. [etc.] Second series, XXXVII, 1864, 212—220.

—— Note on the affinities of the Bellerophontidæ. < Proceedings of the Chicago academy of sciences, I, 9—11, 1865.

—— Check list of the invertebrate fossils of North America. Cretaceous and Jurassic. By F. B. Meek. —— Washington : Smithsonian Institution. April, 1864. [8vo. 1 p. l. 40 pp.—25 c.] < Smithsonian miscellaneous collections. VII, 1867.

MÖRCH (Otto A · · · L · · ·). Review of the Vermetidæ. < Proceedings of the Zoological Society of London for the year 1861, 145—181, pl. 25 (Part I); 326—365 (Part II) ; 1862, 54—83 (Part III).

OWEN (Richard). Mollusca. < The Encyclopædia Britannica, · · · · XV, 1857, 319—403.

[Authority for the subdivision of Tunicates into *Saccobranchiata*, *Dactylobranchiata*, and *Taeniobranchiata*.]

PFEIFFER (Louis). Monographia Pneumonoporum viventium. Sistens descriptiones systematicas et criticas omnium hujus ordinis generum et specierum hodie cognitarum, accedente fossilium enumeratione. · · · , —— Cassellis. Sumptibus Theodori Fischer. 1852. [etc. 8vo. xviii, 439 pp.—2½ th.]

—— Ibid. II. Supplementum primum. · · · · —— Cassellis. Sumptibus Theodori Fischer. 1858. [8vo. viii, 249 pp.—2 th.]

—— Ibid. III. Supplementum secundum. · · · · ——Cassellis Sumptibus Theodori Fischer. 1865. [8vo. 2 p. l. 284 pp.—2½ th.]

—— Catalogue of Phaneropneumona, or terrestrial operculated mollusca, in the collection of the British Museum. —— Printed by order of the trustees. London, 1852. [12mo. 2 p. l. 324 pp.—5 sh.]

[A translation of the Monographia Pneumonoporum viventium (1852), with few modifications, edited by Dr. J. E. Gray.]

—— Monographia Auriculaceorum viventium. Sistens descriptiones systematicas et criticas omnium hujus familiae generum et specierum hodie cognitarum, nec non fossilium enumeratione. Accedente Prosipernaceorum nec non generis Truncatellae historia. Cassellis. Sumptibus Theodori Fischer. 1856. [8vo. xiii, 209 pp.—2 th.]

—— Catalogue of Auriculidæ, Proserpinidæ, and Truncatellidæ in the collection of the British Museum. London : printed by order of the trustees. 1857. [12mo. 2 p. l. 150 pp.—1 sh. 9 d.]

[A translation of the preceding, with slight modifications, edited by Dr. J. E. Gray.]

PHILADELPHIA (Conchological Section of the Academy of Natural Sciences of). [Catalogue of recent Mollusca. Viz:—]

Catalogue of recent Mollusca, belonging to the order Pholadacea. By George W. Tryon, Jr. pp. 1—21. 1868.
Catalogue of the family Solenidæ. By T. A. Conrad. pp. 22—29. 1868.
Catalogue of the family Mactridæ. By T. A. Conrad. pp. 30—47. 1868.
Catalogue of the family Anatinidæ. By T. A. A. Conrad. pp. 49—58. 1869.

Catalogue of the families Saxicaridæ, Myidæ and Corbulidæ. By George W. Tryon, jr. pp. 59—68. 1869.

Catalogue of the family Pandoridæ. By Philip P. Carpenter. pp. 69—71. 1869.

Catalogue of the family Tellinidæ. By George W. Tryon, jr. pp. 72—126. 1869.

Catalogue of the recent species of the family Corbiculadæ. By Temple Prime. pp. 127—187. 1870.

Catalogues of the families Porcellanidæ [=Cypraeidæ+Triviidae—Eratoinæ] and Amphiperasidæ. By S. R. Roberts. pp. 189—214. 1870.

Catalogue of the known species, recent and fossil, of the family Marginellidæ [+ Cystiscidæ + Eratoinæ]. By John II. Redfield. pp. 215—269. 1870.

[Although these catalogues have not actually been referred to in the Arrangement, they are here recorded on account of their usefulness as well as cognate nature.]

See, also, **AMERICAN** Journal of Conchology.

REEVE (Lovell Augustus). Conchologica iconica; or, illustrations of the shells of Molluscous animals. . . . . London : Reeve, brothers, ... 1843 [-] 1845; Reeve, Benham, and Reeve, ... 1847 [-] 1849 ; Reeve and Benham, ... 1851; Lovell Reeve, ... 1854 [-] 1860; Lovell Reeve & co., ... 1862, [et seq.] [4to., 193 monographs in 17 volumes.]

[The following classified list of the "monographs" is given, in order to serve as an index to the volumes—a desideratum that has not been supplied by the publishers—as well as and more especially to serve as a reference from the best known generic names to the position of the families in the present arrangement, and to give some—although rather inadequate—idea of the numbers of species. It must be understood, however, that many of the "genera" enumerated in the following list are artificial assemblages of species combined on account of agreement in some more or less marked conchological character, and that some genera (e. g. *Bulimus, Helix, Lucina, Pyrula,* etc.) contain representatives of several widely distinct families. The references in such cases are to the families containing the typical species of such genera.

The monographs were generally published within a year of dates assigned to the volumes in which they were subsequently combined.

Vo	Year	Pl.	£.	s.	d.	Vol.	Year.	Pl.	£.	s.	d.
1	1843	131	8	10	6	10	1858	126	8	4	0
2	1843	114	7	9	0	11	1859	126	8	4	0
3	1845	130	8	9	0	12	1860	131	8	10	6
4	1847	110	7	4	0	13	1862	126	8	4	0
5	1849	147	9	10	6	14	1864	137	8	18	0
6	1851	129	8	8	0	15	1866	121	8	0	0
7	1854	210	13	15	0	16	1868	127	8	5	6
8	1855	153	9	18	0	17	1870	123			
9	1856	119	7	15	6						

The prices of separate monographs range from 1 sh. 6 d. per plate (1—2 pl.) and 1 sh. 4 d. (3—6 pl.) to little more than 1 sh. 3 d., according to the number of plates.]

CONTENTS.

Class A.—CEPHALOPODA.

Order I.—Dibranchiata.

Order X.—NUDIBRANCHIATA.

No genera monographed.

SUB-CLASS PTEROPODA.

Order XI.—THECOSOMATA.

Order XII.—GYMNOSOMATA.

No genera monographed.

SUB-CLASS PROSOPOCEPHALA.

Order XIII.—SOLENOCONCHA.

No genera monographed.

CLASS C.—CONCHIFERA.

Order XIV.—DIMYARIA.

Order XV.—METARRHIPTÆ.

Order XVI.—HETEROMYARIA.

Order XVII.—MONOMYARIA.

Order XVIII.—RUDISTA.

No genera monographed.

Sub-Branch MOLLUSCOIDEA.

CLASS D.—TUNICATA.

Order XIX.—SACCOBRANCHIA.
Order XX.—DACTYLOBRANCHIA.
Order XXI.—TÆNIOBRANCHIA.
Order XXII.—LARVALIA.

No genera monographed.

CLASS E.—BRACHIOPODA.

Order XXIII.—ARTHROPOMATA.

Order XXIV.—Lyopomata.

STEENSTRUP (Japetus Smith). Overblik over de i Kjöbenhavns museer opbevarede Blæksprutter fra det aabne hav (1860–61). [Cranchiæformes.] < Oversigt over det Kgl. danske viderskabernes selskabs forhandlinger og dets medlemmers arbeider i aaret 1861, 69—86.

STIMPSON (William). On certain genera and families of Zoophagous Gasteropods. < American Journal of Conchology. I, 55—64, pl. 8, 9. 1865.

—— Researches upon the Hydrobiinæ and allied forms; chiefly made upon materials in the museum of the Smithsonian Institution. By Dr. William Stimpson. Washington: Smithsonian Institution. August, 1865. [8vo. 2 p. l. 59 pp.—50 c.] < Smithsonian miscellaneous collections. VII.

TROSCHEL (Franz Hermann). Das gebiss der Schnecken zur begründung einer natürlichen classification untersucht von Dr. F. H. Troschel, · · · ·. Erster band. Mit zwanzig kupfertafeln von Hugo Troschel.——Berlin. Nicolaische verlagsbuchhandlung. (G. Parthey.) 1856—1863. [4to. viii, 252 pp. 20 pl. and 20 l. explan. opposite. Published in 5 parts, lief. 1—4, each 2 th. 20 ngr.; lief. 5, 8 th.; complete, 13¾ th. Zweiten bandes erste [—] dritte lieferung. pp. 1-132, pl. 1-12, 1866—1869; lief. 1—3, each 3 th.]

TRYON (George Washington, jr.). Synopsis of the recent species of Gastrochænidæ [including Brechitidæ], a family of acephalous mollusca. < Proceedings of the Academy of Natural Sciences of Philadelphia. 1861, 465—494.

—— On the classification and synonymy of the recent species of Pholadidæ. < Ib. 1862, 191—221.

—— Monograph of the family Teredidæ. < Ib. 1862, 453—482.

—— Observations on the family Strepomatidæ [=Ceraphasiidæ]. < American Journal of Conchology. I, 1865, 93—135.

—— Monograph of the family Strepomatidæ. < Ib. I, 1865, 299—341; II, 1866, 14—52, 115—133.

TURTON (William). Manual of the land and fresh-water shells of the British islands. With figures of each of the kinds. By William Turton, M. D. New edition, with additions, by John Edward Gray, · · · ·. London: Longman, Brown, Green, Longmans, and Roberts. 1857. (12mo. XVI, 335 pp. 12 pl.)

VAILLANT (Leon). Recherches sur la famille des Tridacnides. < Annales des Sciences Naturelles. Cinquième série. Zoologie et paléontologie. IV, 64—172, pl. 8—12, 1865.

WOODWARD (Samuel P · · ·). A manual of the Mollusca; or, a rudimentary treatise of recent and fossil shells. By S. P. Woodward, A.L.S. · · · Illustrated by A. N. Waterhouse and Joseph Wilson Lowry. London: John Weale, · · ·, MDCCCLI—VI. [12mo. xvi, 486 pp. 1 front. 24 pl. with 12 intercalated leaves explanatory, 1 map.—6 sh. 6 d.—Originally issued in three parts.]

44

WOODWARD (Samuel P···). A manual of the Mollusca: a treatise on recent and fossil shells. By the late S. P. Woodward, A.L.S. [etc.]. With numerous illustrations by A. N. Waterhouse and J. W. Lowry. Second edition. London: Virtue brothers & co., ··· 1866. [12mo. xiv, 518 pp. 1 front. 23 pl. with 12 l. explanatory, 1 map.—5 sh. 6 d.]

—— Appendix to the Manual of the Mollusca of S. P. Woodward, A.L.S., containing such recent and fossil shells as are not mentioned in the second edition of that work. By Ralph Tate, ····.—— London: Virtue & co., ···· 1868. [12mo. 86 pp.—1 sh.]

INDEX TO ARRANGEMENT OF MOLLUSKS.

48

49